Concrete Mixers

Mari Bolte

CREATIVE EDUCATION • CREATIVE PAPERBACKS

seedlings

Published by Creative Education and Creative Paperbacks
P.O. Box 227, Mankato, Minnesota 56002
Creative Education and Creative Paperbacks
are imprints of The Creative Company
www.thecreativecompany.us

Design by Ellen Huber
Production by Alison Derry
Art direction by Tom Morgan

Photographs by iStock (ewg3D, Pixelci, tracielouise), Shutterstock (alexfan32, Baloncici, Bannafarsai_Stock, DmytroPerov, garmoncheg, INDUPHOTOS, KAMONRAT, Nerthuz, Pamela Au, Roman023_photography, sakoat contributor, wanchai, XPhantom)

Library of Congress Cataloging-in-Publication Data
Names: Bolte, Mari, author.
Title: Concrete mixers / Mari Bolte.
Other titles: Seedlings (Creative Education, Inc. (Mankato, Minn.))
Description: Mankato, Minnesota : Creative Education and Creative Paperbacks, [2024] | Series: Seedlings | Includes bibliographical references and index. | Audience: Ages 4–7 | Audience: Grades K–1 | Summary: "An early elementary-level STEM introduction to the concrete mixer, covering how the construction vehicle looks and works. Includes a glossary, further resources, and a labeled image guide to the cement-mixing machine's major parts"— Provided by publisher.
Identifiers: LCCN 2023012496 (print) | LCCN 2023012497 (ebook) | ISBN 9781640267589 (library binding) | ISBN 9781682773086 (paperback) | ISBN 9781640009288 (pdf)
Subjects: LCSH: Concrete mixers—Juvenile literature. | Mixing machinery—Juvenile literature. | Construction equipment—Juvenile literature. | CYAC: Concrete mixers. | Mixing machinery. | Construction equipment. | LCGFT: Instructional and educational works.
Classification: LCC TA439 .B577 2024 (print) | LCC TA439 (ebook) | DDC 620.1/360284—dc23/eng/20230330
LC record available at https://lccn.loc.gov/2023012496
LC ebook record available at https://lccn.loc.gov/2023012497

Printed in China

TABLE OF CONTENTS

Time to Mix! 4

Big Trucks 6

What's Inside 8

Spinning Drums 10

Heavy Loads 12

Get Moving 14

Down the Slide 16

All Done Mixing! 18

Picture a Concrete Mixer 20

Words to Know 22

Read More 23

Websites 23

Index 24

Time to mix!

Concrete mixers are big trucks.

They mix cement and water together.

A large drum sits on the back of the truck. Concrete is mixed inside the drum.

A motor makes the drum spin. Spinning keeps the wet concrete from hardening.

Concrete is heavy! Mixers are powerful trucks. They can carry twice their weight in concrete.

Mixer trucks move on strong wheels. They bring concrete to building sites.

Mixer trucks have a chute. It is often on the back. Concrete runs out and down the chute.

All done mixing!

Picture a Concrete Mixer

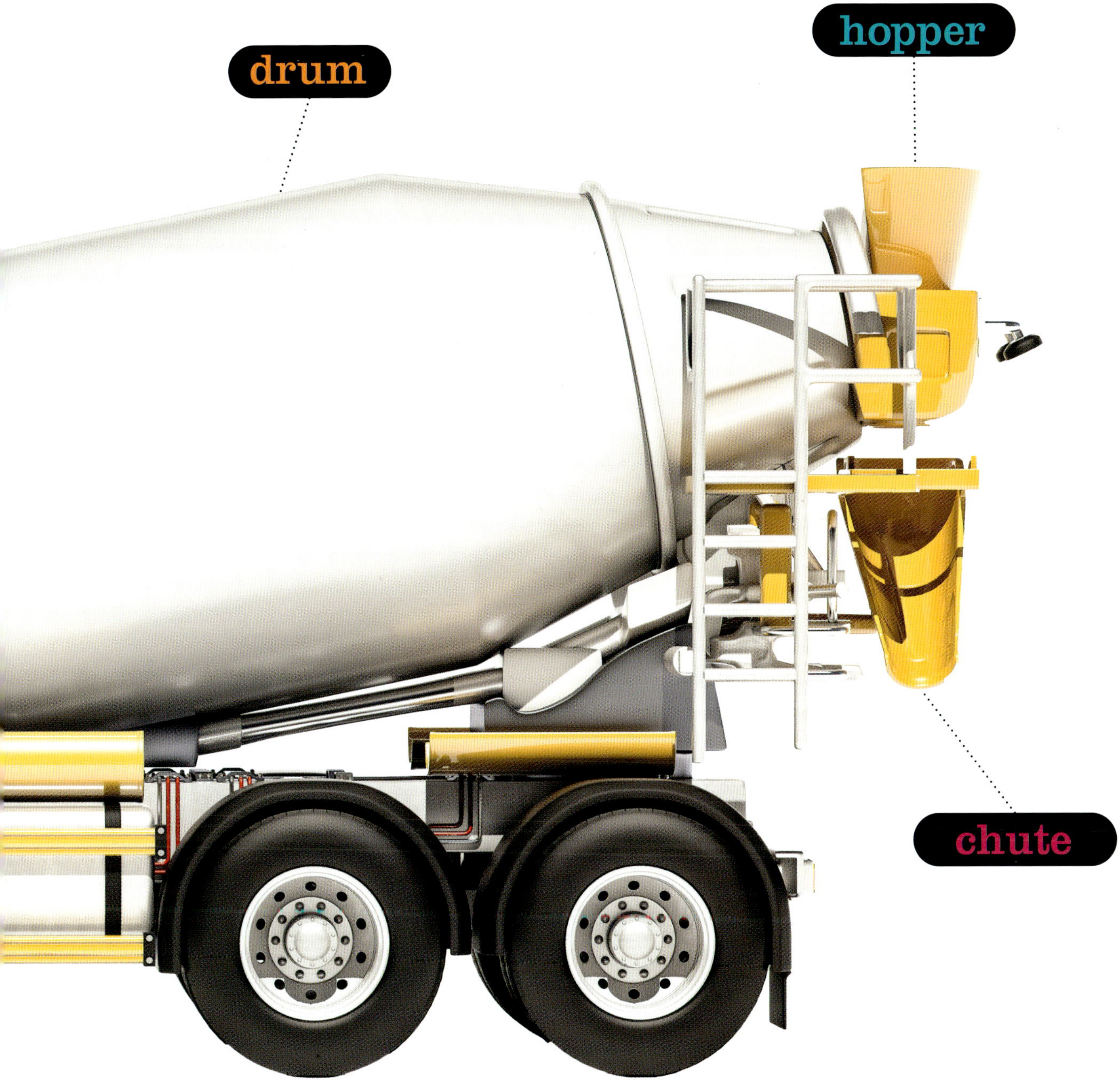
drum
hopper
chute

Words to Know

cement: a powder made from clay and rock

chute: a slide

concrete: a building material made from cement and water

drum: a spinning container

Read More

Cella, Clara. *Underwater Construction Workers.* Minneapolis: Lerner Publications, 2023.

Humphrey, Natalie. *Incredible Earthmovers.* New York: Gareth Stevens Publishing, 2023.

Websites

Cement
https://kids.britannica.com/kids/article/cement/400093
Learn what cement is and how it is used.

Cement Facts for Kids
https://kids.kiddle.co/Cement
Cement is everywhere—find out exactly where!

Index

cement, 7
chutes, 17
drums, 8, 10
mixing, 4, 7, 8, 18
motors, 10
spinning, 10
water, 7
weight, 13
wheels, 14